Pocketbook of Environmental/Safety Compliance—*Abbreviation, Acronyms, Elements of Calculation, Regulations and Websites*

By

Satya R. Chatterjee, Ph.D., P.E.

SRC Technologies Inc., Houston, Texas

ISBN: 1-4140-0287-4 (e-book)
ISBN: 1-4140-0300-5 (Paperback)

This book is printed on acid free paper.

1stBooks - rev. 06/23/04

FOREWORD

The purpose of this pocketbook is to introduce the readers with the firsthand information sources about Environment/Safety and then connect them with EPA, OSHA, pollution concerns, toxic wastes and other environmental issues. Research for useful information from vast sources to the readers in an organized simple form has convinced the author that there is a vast ocean between the resources and the readers for improved and effective communication. The intent of the pocketbook is to narrow this gap.

Although the author took pain in preparing this pocketbook so that the contents are accurate and correct, in any case, he, himself or anybody associated with this publication does not commit with any warranty, expressed or implied, in providing the information.

The environmental field is vast, safety issues are complex, compliance dilemmas are challenging. In this realm, if the readers can find the source of advancing their information resulting from the use of this pocketbook, the author's endeavor shall be rewarding.

This pocketbook has been designed for everyone to help everyone as a practical information source.

Credit must be expressed for EPA, OSHA and other professional organizations whose research and design information are our guiding values for environmental protection and compliance.

CONTENTS

APPENDICES

SELECTED ABBREVIATIONS

A

Alternating Current	ac
Amount of Substance, mole	mol
Area, hectare	ha
Area, square meter	m^2
Acceleration	m/s^2

B

Barrel	bbl
British thermal unit	Btu

C

Celsius (**temperature**)	C
Centimeter	cm
Cubic foot	ft^3
Cubic meter	m^3

D

Decibel	dB
Decibel on A weighted scale	dBA
Density	kg/m^3
Diameter	dia
Direct Current	dc

E

Electric Current, (**ampere**)	A
Electron Volt	eV
Energy or Work, joule	J

F

Fahrenheit	F
Feet per second	fps
Foot (feet)	ft
Force, Newton	N

G

Gallon	gal
Gallons per day	gpd
Grains per ounce	gr/oz
Grams per gallon	GPG

H

Hectare	ha
Hertz (frequency)	Hz
Hour	h

I

Inch	in.
Inside Diameter	ID

J

Joule (energy)	J

K

Kelvin	K
Kilogram	kg
Kilometer	km
Kilowatt	kw
Kilowatt hour	kwh

L

Length, (**meter**)	m
Length, (**Kilometer**)	km
Liter	l

M

Mass or weight, (**gram**)	g
Mass or weight, (**kilogram**)	kg
Meter	m
Megahertz	MHz
Megawatt	MW
Microgram	µg
Mile	mi
Milligram	mg
Millihertz	mHz
Milliliter	ml
Millisiemens per meter	mS/m

N

Nautical Mile	NM

O

Ohm	Ω
Ohm-meter	Ωm
Outside Diameter	OD
Ounce	oz

P

Parts per billion, ppb	µg/l
Parts per million, ppm	mg/l, mg/m³, ml/m³
Pound	lb
Pounds per square inch (Press.)	psi
Pounds per square inch (gauge)	psig
Power, watt	W
Pressure, (**bar or pascal**)	bar or Pa

Q

Quadrillion Btu	qBtu

R

Revolutions per minute	rpm

S

Siemens, conductance	S
Specific Gravity	Sp.Gr.
Surface Tension	N/m

T

Temperature, (**Kelvin-absolute**)	K
Temperature, (**degree Celsius**)	°C
Time, (**second**)	s
Tons per day	TPD

V

Velocity	m/s
Viscosity—dynamic, poise	P
Viscosity—kinematic, stoke	St
Volume, liter	l
Volt	V

W

Watt (power)	W
Watt-hours	Wh

ACRONYMS

A

AA	Atomic Absorption; Atomic Absorption Spectroscopy
AAA	American Arbitration Association
AAAS	American Association for the Advancement of Science
AAEE	American Academy of Environmental Engineers
AAES	American Association of Engineering Societies
AAP	Affirmative Action Plan; Asbestos Action Program
ABA	American Bar Association
ABMA	American Boiler Manufacturers Association
ACA	American Conservation Association
ACC	American Chemistry Council
ACE	Alliance for Clean Energy
ACEC	American Consulting Engineers Council
ACGIH	American Conference of Government Industrial Hygienists
ACL	Analytical Chemical Laboratory; Air Contaminants Limit
ACM	Asbestos Containing Material
ACP	Area Contingency Plan
ACS	American Chemical Society;

ACS	American Cancer Society
ACWA	American Clean Water Association
ADB	Applications Data Base
AEA	Atomic Energy Act
AEE	Alliance of Environmental Education
AEHA	Army Environmental Hygiene Agency (U.S.)
AES	Auger Electron Spectrometry
AFA	American Forestry Association
AGA	American Gas Association
AGI	American Geological Institute
AHERA	Asbestos Hazard Emergency Response Act
AHM	Acutely Hazardous Material
AHW	Acutely Hazardous Waste
AI	Artificial Intelligence
AIA	American Institute of Architects; Asbestos Information Association
AIChE	American Institute of Chemical Engineers
AIC	American Institute of Chemists
AIHA	American Industrial Hygiene Association
AISI	American Iron and Steel Institute
AMA	American Medical Association
AMC	American Mining Congress; Airborne Molecular Contaminants
AMS	American Meteorological Society
AMSA	Association of Metropolitan Sewerage Agencies
ANEC	American Nuclear Council

ANSI	American National Standards Institute
AOC	Abnormal Operating Conditions
APA	Administrative Procedure Act; Acid Precipitation Act; American Planning Association
APCA	Air Pollution Control Association
API	American Petroleum Institute
APPA	American Public Power Association
APWA	American Public Works Association
AQD	Air Quality Digest
AQIA	Air Quality Impact Assessment
AQMP	Air Quality Management Plan
ARB	Air Resources Board
ARI	Air-Conditioning and Refrigeration Institute
ARIP	Accidental Release Information Program
ARM	Air Resources Management
ARRPA	Air Resources Regional Pollution Assessment
AS	Air Sparging; Area Source
ASCE	American Society of Civil Engineers
ASCII	American Standard Code for Information Interchange
ASCP	American Society of Consulting Planners
ASHAA	Asbestos in Schools Hazard Abatement Act

ASHRAE	American Society for Heating, Refrigeration and Air Conditioning Engineers
ASME	American Society of Mechanical Engineers
ASPA	American Society of Public Administration
ASPIS	Abandoned Site Program Information System
ASPM	Abnormal Situation Prevention and Management
ASQC	American Society of Quality Control
ASSE	American Society of Sanitary Engineers; American Society of Safety Engineers
AST	Aboveground Storage Tank
ASTM	American Society for Testing and Materials
ASU	Activated Sludge Unit
AT	Advanced Treatment
ATB	Asphalt Treated Base/Permeable Base
ATMI	American Textile Manufacturing Institute
ATSDR	Agency for Toxic Substances and Disease Registry
ATTRA	Appropriate Technology Transfer for Rural Areas
AVGAS	Aviation Gasoline

AWSE — Association of Women in Science and Engineering
AWMA — Air and Waste Management Association
AWPI — American Wood Preservers Institute
AWQC — Ambient Water Quality Criteria
AWRA — American Water Resources Association
AWWA — American Water Works Association
AWWARF — American Water Works Association Research Foundation

B

BACT — Best Available Control Technology
BAT — Best Available Technology; Best Available Treatment
Bbl — Barrel
BCF — Bio-concentration Factor
BCT — Best Conventional Control Technology
BD — Business Development
BDT — Best Demonstrated Technology
BEJ — Best Engineering Judgment; Best Expert Judgment
BEST — Best Extractive Sludge Treatment
BHC — Benzene Hexa-chloride
BHP — Brake Horsepower
BIF — Boiler and Industrial Furnace
BLM — Bureau of Land Management

BLS	Bureau of Labor Statistics
BMP	Best Management Practices; Best maintenance Practice
BNA	Bureau of National Affairs
BOD5	Biochemical Oxygen Demand, 5-day
BOF	Basic Oxygen Furnace
BOM	Bureau of Mines
BOT	Build, Operate (or Own), Transfer
BPB	Biological Permeable Barrier
BPT	Best Practicable Control Technology; Best Practicable Treatment
BRA	Baseline Risk Assessment
BS	British Standard
BSI	British Standard Institute
BSO	Benzene Soluble Organics
BTEX	Benzene, Toluene, Ethyl-benzene and Para-xylenes— Volatile Organic Compounds
Btu	British Thermal Unit
BY	Budget Year

C

CA	Corrective Action; Cost Analysis
CAA	Clean Air Act
CAAA	Clean Air Act Amendments
CAD	Computer Aided Design

CADD	Computer Aided Drafting and Design
CAM	Compliance Assurance Monitoring; Continuous Air Monitoring Program
CAP	Capacity Assurance Program; Corrective Action Plan; Cost allocation Procedure
CAR	Corrective Action Report
CAS	Chemical Abstracts Service
CAU	Carbon Adsorption Unit
CB	Continuous Bubbler
CBA	Cost Benefit Analysis
CC	Carbon Copy
CCAP	Center for Clean Air Policy
CCHW	Citizens Clearinghouse for Hazardous Wastes
CCL	Compacted Clay Layer
CCP	Composite Correction Plan
CDC	Centers for Disease Control (US)
CDM	Comprehensive Data Management; Climate-logical Dispersion Model
CEA	Council of Economic Advisors
CEB	Chemical Element Balance
CEC	Cation Exchange Capacity; Commission of European Communities; Consulting Engineers Council
CEI	Compliance Evaluation Inspection; Competitive Enterprise Institute
CEM	Continuous Emission Monitoring

CEMS	Continuous Emission Monitoring System
CEO	Chief Executive Officer
CEQ	Council on Environmental Quality
CERCLA	Comprehensive Environmental Response, Compensation And Liability Act
CF	Conservation Foundation; Conversion Factor
CFC	Chloro-fluoro-methanes
CFR	Code of Federal Regulations
CGI	Combustible Gas Indicator
CGL	Comprehensive General Liability
CGP	Construction General Permit
CHP	Chemical Hygiene Plan
CHIP	Chemical hazard Information Profile
CIAQ	Council on Indoor Quality
CIH	Certified Industrial Hygienist
CII	Construction Inspection Institute
CIMS	Chemical Information Management Systems
CIS	Chemical Identification System; Chemical Information System
CLF	Conservation Law Foundation
CM	Corrective Measure
CMA	Chemical Manufacturers Association
CMB	Chemical Mass Balance
CME	Comprehensive Monitoring Evaluation

CMMS	Computerized Maintenance Management System
CMTL	Construction materials Testing Laboratory
CNG	Compressed Natural Gas
CNS	Central Nervous System
CO	Carbon Monoxide; Change Order
CO2	Carbon Dioxide
COD	Chemical Oxygen Demand
COE	Corps of Engineers (U.S. Army)
CPA	Certified Public Accountant; Cost Plus Award
CPC	Chemical protective Clothing
CPF	Cost Plus Fixed; Carcinogenic Potency Factor
CPI	Chemical Process Industries; Consumer Price Index; Cost Plus Incentive
CPM	Cycles per Minute
CPR	Cardiopulmonary Resuscitation; Cost Performance Report
CPSA	Consumer Product Safety Act
CPSC	Consumer Product Safety Commission
CPVC	Chlorinated Polyvinyl Chloride
CQA	Construction Quality Assurance
CRP	Conservation Reserve Program; Community Relation Program
CRT	Cathode Ray Tube

CSB	Chemical Safety and Hazard Investigation Board
CSEP	Confined Space Entry Program
CSI	Compliance Sampling Inspection Common Sense Initiative (U.S. EPA)
CSO	Combined Sewer Overflow
CSP	Conservation Security Program Certified Safety Professional
CSPI	Center for Science in the Public Interest
CTB	Cement Treated Base
CVR	Closure Verification Report
CWA	Clean Water Act
CWRT	Center for Waste Reduction Technologies
CWTC	Chemical Waste Transportation Council

D

DAF	Department Of Air Force
DCO	Document Control Officer
DCS	Distributed Control System
DDD	Dichloro-diphenyl-dichloroethane
DDE	Dichloro-diphenyl-dichloroethylene
DDT	Dichloro-diphenyltrichloroethane
DHS	Designated Hazardous Substances
DI	Diagnostic Inspection
DMP	Data Management Plan

DMS	Data Management System
DN	Department of Navy
DNA	Deoxyribonucleic Acid
DNB	Dinitrobenzene
DNT	Dinitrotoluene
DO	Dissolved Oxygen
DOA	Department of Agriculture (USDA)
DOC	Department of Commerce
DOD	Department of Defense
DOE	Department of Energy
DOHHS	Department of Health and Human Services
DOHS	Department of Homeland Security
DOI	Department of Interior
DOJ	Department of Justice
DOL	Department of Labor
DOS	Department of States
DOT	Department of Transportation
DPDT	Double Pole Double Throw
DPST	Double Pole Single Throw
DWS	Drinking Water Standard

E

EA	Environmental Assessment (Action, Audit); Exposure Assessment
EAP	Environmental Action Plan
E&P	Exploration and Production
EB	Emission Balancing

EBS	Environmental Baseline Survey
ECL	Environmental Chemical Laboratory
ED	Effective Dose
EDB	Ethylene Dibromide
EDC	Ethylene Dichloride
EDF	Environmental Defense Fund
EDTA	Ethylene Diamine Triacetic Acid
EE	Environmental Evaluation; Engineering Evaluation
EEG	Electroencephalogram
EER	Excess Emission Report
EF	Emission Factor
EGR	Exhaust Gas Recirculation
EHS	Extremely Hazardous Substances; Environmental Health Science; Environment, Health and Safety
EI	Emissions Inventory
EIA	Environmental Impact Assessment
EIR	Environmental Impact Report
EIS	Environmental Impact Statement; Environmental Inventory System; Emissions Inventory System
EL	Exposure Level; Elastic Limit
EM	Electron Microscope
E-Mail	Electronic Mail
EMAP	Environmental Monitoring and Assessment Program
EMR	Environmental Management Report
EMS	Environmental Management System
EMTS	Exposure Monitoring Test Site

EO	Ethylene Oxide; Executive Officer (Order)
EOC	Emergency Operating Center
EOE	Equal Opportunity Employer
EPA	Environmental Protection Agency (US)
EPAA	Environmental Programs Assistance Act
EPCRA	Emergency Planning and Community Right-to-know Act
EPR	Ethylenepropylene Rubber
EPRI	Electric Power Research Institute
EQIP	Environmental Quality Incentives Program
ERA	Ecological Risk Assessment; Environmental Risk Assessment
ERC	Emergency Response Commission; Emission Reduction Credit; Environmental Research Center
ERCS	Emergency Response Cleanup Services
ERL	Environmental research Laboratory
ESA	Endangered Species Act; Environmentally Sensitive Area
ESH	Environmental Safety and Health
ESP	Electrostatic Precipitator
ETP	Emission Trading Policy
EWS	Engineering Work Station

F

FAA	Federal Aviation Administration
FACA	Federal Advisory Committee Act
FBC	Fluidized Bed Combustion
FCC	Federal Communication Commission; Fluid Catalytic Converter
FCCU	Fluid Catalytic Cracking Unit
FDA	Food and Drug Administration
FE	Flow Element; Fugitive Emissions
FEA	Federal Energy Administration
FEMA	Federal Emergency Management Administration
FEPCA	Federal Environmental Pesticide Control Act
FERC	Federal Energy Regulatory Commission
FFA	Flammable Fabrics Act
FFDCA	Federal Food, Drug and Cosmetic Act
FGD	Flue Gas Desulfurization
FHA	Federal Housing Authority
FHSA	Federal Hazardous Substances Act
FIA	Federal Insurance Administration
FID	Flame Ionization Detector
FIFRA	Federal Insecticide, Fungicide and Rodenticide Act
FIPS	Federal Information Processing Standards

FOIA	Freedom of Information Act
FP	Flash Point
FPA	Federal Pesticide Act
FPC	Federal Power Commission
FPD	Federal Photometric Detector
FPP	Farmland Protection Program
FR	Federal Register
FRA	Federal Register Act
FRP	Fiberglass Reinforced Plastic
FS	Feasibility Study; Full Scale
FTC	Federal Trade Commission
FTU	Fixed Treatment Unit
FWCA	Fish and Wildlife Coordination Act
FWPCA	Federal Water Pollution Control Administration
FY	Fiscal (Financial) Year

G

G	Gibbs (free energy)
GAC	Granular Activated Carbon
GACT	Generally Available Control Technology; Granular Activated Carbon Treatment
GAO	General Accounting Office
GC	Gas Chromatography
GDP	Gross Domestic Product

GEMS	Global Environmental Monitoring System
GEP	Good Engineering Practice
GHG	Greenhouse Gas
GIS	Global Indexing System
GLC	Gas Liquid Chromatography
GMS	Groundwater Modeling System
GP	General Permits
GRI	Gas Research Institute
GS	Geological Survey (U.S.)
GSA	General Services Administration
GWPS	Groundwater Protection Standard

H

H	Humidity; Henry's Law Constant
H2O	Water
H2S	Hydrogen Sulfide
HA	Hazard Assessment
HACCP	Hazard Analysis for Critical Control Points
HAP	Hazardous Air Pollutants
HAZCOMM	Hazardous Communication Program
HAZMAT	Hazardous Material
HAZOP	Hazard and Operability Study
HAZWOPER	Hazardous Waste Operations and Emergency Response
HC	Hydrocarbon; Hazardous Chemicals;

HC	Hazardous Constituents
HCS	Hazard Communication Standard
HDPE	High Density Polyethylene
HHC	Highly Hazardous Chemicals
HHV	High Heating Value
HI	Hazard Index
HMR	Hazardous Materials Regulations
HMTA	Hazardous Materials Transportation Act
HOC	Hazardous Organic Constituents
HOV	High Occupancy Vehicle
HP	Horse Power
HRS	Hazard Ranking System
HSL	Hazardous Substances List
HSP	Health and Safety Plan
HTRW	Hazardous, Toxic and Radioactive Waste
HUD	Housing and Urban Development (U.S)
HVAC	Heating, Ventilation and Air Conditioning
HWCL	Hazardous Waste Control Law
HWM	Hazardous Waste Management
HWTC	Hazardous Waste Treatment Council

I

IAEA	International Atomic Energy Agency
IBA	Industrial Biotechnology Association
ICC	Interstate Commerce Commission; International Code Council
ICRP	International Commission on Radiological Protection
ID	Inside Diameter
IEB	International Environmental Bureau
IEC	International Electrotechnical Commission
IEEE	Institute of Electrical and Electronic Engineers
IFB	Invitation for Bid
IFCI	International Fire Code Institute
IG	Industrial Hygienist
IOU	Input Output Unit
IR	Infra-red
IRPTC	International Register for Potentially Toxic Chemicals
IRS	Internal Revenue Service
ISO	International Organization for Standardization
ISA	Instrument Society of America; Instrumentation Systems and Automation
ITC	International Trade Commission
IWA	Inside Work Area

J

JAPCA	Journal of Air Pollution Control Association
JEEP	Joint Emissions Estimation Program

K

KMNO4	Potassium Permanganate
KW	Kilowatts

L

LAER	Lowest Achievable Emission Rate
LC	Lethal Concentration; Liquid Chromatography
LCL	Lower Control Limit
LD	Lethal Dose
LDAR	Leak Detection and Repair
LDPE	Low Density Polyethylene
LEL	Lower Explosive Limit
LEPC	Local Emergency Planning Committee
LHW	Liquid Hazardous Waste
LOPA	Layers of Protection Analysis
LNG	Liquefied Natural Gas
LPG	Liquefied Petroleum Gas
LUFT	Leaking Underground Fuel Tank
LUST	Leaking Underground Storage tank
LWL	Lower Warning Limit

M

MACT	Maximum Achievable Control Technology
MCA	Manufacturing Chemists Association
MCL	Maximum Contaminant Level
MCRT	Mean Cell Residence Time
MH	Man-Hour (Hole)
MIC	Methyl Isocyanate
MIS	Management Information System
MIT	Mechanical Integrity Level
MLSS	Mixed Liquor Suspended Solids
MLVSS	Mixed Liquor Volatile Suspended Solids
MMSCFD	Million Standard Cubic feet Per Day
MOA	Memorandum of Agreement
MOC	Management of Change
MOI	Memorandum of Intent
MOU	Memorandum of Understanding
MPH	Miles Per Hour
MPN	Most Probable Number
MREM	Milliroentgen Equivalent in Man
MS	Mass Spectrometry
MSDS	Material Safety Data Sheet
MTBF	Mean Time Between Failures
MTC	Metropolitan Transit (Transportation) Commission
MTD	Maximum Tolerated Dose
MW	Molecular Weight
MWTA	Medical Waste Treatment Act

N

N	Nitrogen
NAAQS	National Ambient Air Quality Standards
NAC	National Asbestos Council
NACA	National Agricultural Chemicals Association
NAE	National Academy of Engineering
NALGEP	National Association of Local Govt. Env. Professionals
NAM	National Association of Manufacturers
NAR	National Asbestos Registry
NARA	National Air Resources Act; National Archives and Record Administration
NAS	National Academy of Sciences
NASA	National Aeronautics and Space Administration
NAWDEX	National Water Data Exchange
NBP	National Bio-solids Partnership
NCA	Noise Control Act
NCAF	National Clean Air Fund
NCAQ	National Commission on Air Quality
NCI	National Cancer Institute
NCP	National Contingency Plan
NCSE	National Council of Sciences on Environment

NDT	Nondestructive Testing
NEA	National Energy Act
NEC	National Electric Code
NEMA	National Electric Manufacturers Association
NEPA	National Environmental Policy Act
NESHAP	National Emission Standards for HAPs
NFPA	National Fire Protection Association
NGPA	Natural Gas Policy Act
NIEHS	National Institute of Environmental Health Sciences
NIH	National Institutes of Health
NIOSH	National Institute of Occupational Safety and Health
NIR	Near Infra-red
NIST	National Institute of Standards and Technology
NLE	National Library of Environment
NLM	National Library of Medicine
NMP	National Municipal Policy
NO	Nitric Oxide
NO2	Nitrogen Dioxide
NOI	Notice of Intent
NOT	Notice of Termination
NOX	Oxides of Nitrogen
NPDES	National Pollution Discharge Elimination System (Permit)
NPS	Non-point Source
NRC	National Research Council; Nuclear Regulatory Commission

NRDC Natural Resources Defense Council
NSC National Safety Council
NSF National Sanitation Foundation; National Science Foundation
NSPE National Society of Professional Engineers
NSPS New Source Performance Standards
NSR New Source Review
NTSB National Transportation Safety Board
NWF National Wildlife Federation
NWPA Nuclear Waste Policy Act
NWS National Weather Service

O

O2 Oxygen
O3 Ozone
OCS Outer Continental Shelf
OD Outside Diameter; Organizational Development
OEM Original Equipment Manufacturer
OMB Office of Management and Budget
OPA Oil Pollution Act
ORP Oxidation-Reduction Potential
OSHA Occupational Health and Safety Administration (U.S.); Occupational Safety and Health Act
OVA Organic Vapor Analyzer

P

PA	Policy Analyst
PACT	Powdered Activated Carbon Treatment
PASS	Permit Application Software System
Pb	Lead
PCA	Portland Cement Association
PCB	Polychlorinated Biphenyl
PD	Positive Displacement
PE	Performance Evaluation; Professional Engineer
PEL	Permissible Exposure Limit
PFD	Probability of Failure on Demand; Process Flow Diagram
P&ID	Piping and Instrument Diagram
PL	Public Law
PLC	Programmable Logic Controller
PM 2.5	Particulate Matters with dia. of 2.5 microns or less
PM 10	Particulate Matters with dia. of 10 microns or less
POM	Particulate Organic Matter
POTW	Publicly Owned Treatment Works
PPE	Personal Protective Equipment
PPP	Pollution Prevention Plan
PRV	Pressure Relief Valve
PS	Point Source; Pump Station

PSA Pipeline Safety Act
PSD Prevention of Significant Deterioration of Air Quality
PSM Process Safety Management (OSHA);
Point Source Monitoring
PSNS Pretreatment Standards for New Sources
PTFE Poly-tetra-fluoro-ethylene
PUC Public Utility Commission
PVC Polyvinyl Chloride
PWA Present Worth Analysis
PWS Public Water System (Supply)

Q

QA Quality Assurance
QC Quality Control
QCI Quality Control Index

R

RA Regulatory Analysis;
Remedial Action;
Risk Analysis (Assessment)
RACT Reasonably Available Control Technology
RAD Radiation
RAP Radon Action Plan

RBC	Red Blood Count
RCRA	Resource Conservation and Recovery Act
R&D	Research and Development
REL	Recommended Exposure Limit
RF	Response Factor
RH	Relative Humidity
RHW	Restricted Hazardous Waste
RIC	Radon Information Center
RMP	Risk Management Plan (Programs) (EPA)
RO	Reverse Osmosis
RQ	Reportable Quantity
RR	Respiration Rate
RRF	Risk Reduction Factor (RRF = 1/PFD)
RRT	Relative Retention Time
RSP	Respirable Suspended Particles
RT	Retention Time
RTD	Resistance Temperature Detector
RV	Residual Volume
RVP	Reid Vapor Pressure

S

S	Solubility
SA	Sample Analysis; Sludge Age
SAP	Sampling and Analysis Plan; Scientific Advisory Panel

SARA	Superfund Amendments and Reauthorization Act
SCADA	Supervisory Control and Data Acquisition
SDWA	Safe Drinking Water Act
SEC	Securities and Exchange Commission
SEP	Standard Engineering Practice
SFPE	Society of Fire Protection Engineers
SG	Specific Gravity
SI	Site Inspection
SIC	State Industrial Classification
SIL	Safety Integrity Level (SIL 1, SIL 2, SIL 3, SIL 4)
SIP	State Implementation Plan
SIS	Safety Instrumented Systems
SLA	Submerged Lands Act
SO2	Sulfur Dioxide
SOP	Standard Operating Procedure
SOR	Standard Oxygen Required
SOW	Scope of Work
SP	Set Point
SPC	Statistical Process Control
SPCC	Spill Prevention, Control and Countermeasures
SSO	Sanitary Sewer Overflows
SS	Sanitary Sewers; Superfund Surcharge; Suspended Solids
SSM	Startup, Shutdown and Malfunction
SSP	Site Safety Plan

SRT	Solids Retention Time
STEL	Short-term Exposure Limit
STP	Sewage Treatment Plant; Standard Temperature and Pressure
SWDA	Solid Wastes Disposal Act
SWE	Society of Women Engineers
SWMP	Storm Water Management Plan
SWP3	Storm Water Pollution Prevention Plan
SWTCP	Surface Water Toxic Control program

T

TCLP	Toxicity Characteristic Leaching Procedure
TD	Toxic Dose
TDS	Total Dissolved Solids
TFE	Tetrafluoroethylene
TIP	Total Ionization Probe
TKN	Total Kjeldahl Nitrogen
TLV	Threshold Limit Value
TMDL	Total Maximum Daily Load
TOC	Total Organic Carbon
TP	Toxic Pollutants
TPD	Tons per Day
TPQ	Threshold Planning Quantity
TPY	Tons per Year
TQM	Total Quality Management

TNT	Tri-Nitro Toluene
TRI	Toxic Release Inventory
TS	Total Solids; Toxic Substances
TSCA	Toxic Substances Control Act
TSD	Treatment, Storage and Disposal
TSS	Total Suspended Solids; Ten States Standards
TTO	Total Toxic Organics
TVH	Total Volatile hydrocarbons
TWA	Time Weighted Average

U

UBC	Uniform Building Code
UCL	Upper Control Limit
UEL	Upper Explosive Limit
UFC	Uniform Fire Code
UFL	Upper Flammable Limit
UIC	Underground Injection Control
UL	Underwriters Laboratories
ULEV	Ultra Low Emission Vehicle
UN	United Nations
UNEP	United Nations Environment Program
UPV	Unfired Pressure Vessel
URL	Uniform Resource Locators (Web Addresses)
USC	U.S. Code
USCG	U.S. Coast Guard

USDW	Underground Source of Drinking Water
USPS	United States Postal Service
UST	Underground Storage Tank
UV	Ultra-violet
UWL	Upper Warning Limit

V

VCM	Vinyl Chloride Monomer
VES	Vapor Extraction System
VHAP	Volatile Hazardous Air Pollutant
VOC	Volatile Organic Compound
VOD	Valve Operating Diagram
VP	Vapor Pressure
VSS	Volatile Suspended Solids

W

WAP	Waste Analysis Plan
WAS	Waste Activated Sludge
WEF	Water Environment Federation
WE&T	Water Environment and Technology
WHIP	Wildlife Habitat Incentives Program
WHO	World Health Organization
WIN	Water Infrastructure Network
WMMA	Waste Materials Management Act
WP	Work Plan
WRDA	Water Resources Development Act

WRI World Resources Institute
WRP Wetlands Reserve Program
WSO World Safety Organization
WQA Water Quality Association
WQFA Water Quality Financing Act
WQMP Water Quality Management Plan
WTP Water Treatment Plant
WWTP Waste Water treatment Plant
WWW World Wide Web

Y

YTD Year to Date

Z

ZEL Zero Emission Vehicle
ZRL Zero Risk Level

MAJOR ENVIRONMENTAL/SAFETY REGULATIONS

1. **Clean Air Act:**

   The clean air act (CAA) regulates through audit, inspection and operating permits, control of air contaminants from combustion sources, fugitive emissions and sulfur related plant emissions etc.

2. **Clean Water Act:**

   The clean water act (CWA) regulates through various means and counter measures any release of oil into waters of the Unites States.

3. **Oil Pollution Act:**

   The oil pollution act of 1990 regulates through preparation, submission and approval of a response plan for CWA.

4. **Safe Drinking Water Act:**

The safe drinking water act (SDWA) controls through different means the water systems serving the public or any organization being served above certain number of people.

5. **Toxic Substance Control Act:**

The toxic substance control act (TSCA) identifies and controls the chemical products that are considered as risk to humans and the environment.

6. **Resource Conservation and Recovery Act**:

The resource conservation and recovery act (RCRA) controls and monitors the disposal and storage of the solid and recyclable wastes.

7. **Comprehensive Environmental Response, Compensation and Liability Act:**

   The comprehensive environmental response, compensation and liability act (CERCLA) provides requirements for clean-up and funds for hazardous waste sites.

8. **Hazard Communication Standard:**

   The hazard communication standard (HAZCOMM) is a requirement of OSHA mainly for the employers. This program requires that any employer exposing any employee to any quantity of a hazardous chemical shall:

   Prepare a written HAZCOMM program,
   Train employees in order to be knowledgeable of the program,
   Prepare a list of the hazardous Chemicals in the working place,
   Maintain a latest MSDS for those Chemicals and
   Label all hazardous Chemicals in the working place

9. **Superfund Amendments and Reauthorization Act:**

The superfund amendments and reauthorization act (SARA) requires that the state, local agency and the fire department be notified of the presence of 10,000 pounds or more of a hazardous chemical or an extremely hazardous substance. SARA also requires to turn in an annual inventory of these materials on or before March 1 of each year.

10. **Emergency Planning and Community Right-to-know Act:**

This act was enacted mandating the establishment of emergency planning districts and local emergency planning committees along with the publication of TRI data.

Along with these acts, information related to the Endangered Species Act, National Historic Preservation Act and other related provisions are required for compliance permits.

COMMON ENVIRONMENTAL HAZARDS

The common home/environmental hazards are mainly water contamination, radon, pesticides, lead, asbestos and other indoor air pollutants besides nearby hazardous waste sites including nuclear and/or weapon plant sites and outside air contamination.

1. There are four common categories of water contaminants. They are:
   Microbiological contaminants, Inorganics, Organics and Radionuclides. Water must be drinkable per EPA standard. EPA has a basic primary drinking water standard and a secondary standard for variables like color, smell or taste etc.

2. Radon is a naturally occurring inert radioactive uranium gas formed in the decaying process of uranium.

3. Pesticides are chemical or group of chemicals marketed for killing pests like insects, fungi etc.

4. Lead is a heavy, soft, malleable, bluish gray metal which can form toxin if adsorbed by human.

5. Asbestos is a naturally occurring mineral fiber mainly found in rocks. Asbestos is processed to make products used in building materials, pipe insulation, fireboards and plastic products.

6. Indoor air pollutants are contaminated gases or airborne particles from any source that can make indoor air of a house or a building unpleasant.

7. Hazardous wastes are primarily industrial chemicals and municipal wastes like garbage sources. There are thousands of these sites which may create serious health and environmental problems in the country.

8. Nuclear power reactor sites and weapons plant sites along with their waste disposal sites are also concern for health and environmental safety.

9. Outside air are contaminated mainly by automobile and industrial/power plant flare stack emissions.

COMMON TOXIC POLLUTANTS

EPA listed these toxic pollutants each having its own threshold level. Each element may have different threshold level than its compounds.

Acenaphthalene	Acrolein
Acrylonitrile	Aldrin
Antimony	Arsenic
Asbestos	Benzene
Benzidine	Beryllium
Cadmium	Carbontetrachloride
Chlorodane	Chlorinated benzenes
Chlorinated ethanes	Chloroalkyl ethers
Chlorinated naphthalenes	Chlorinated phenols
Chloroform	Chlorophenol
Chromium	Copper
Cyanides	DDT

Dichlorobenzenes	Dichlorobenzedine
Dichloroethylenes	Dichlorophenol
Dichloropropane	Dimethylphenol
Dinitrotoluene	Diphenylhydrazine
Endosulfan	Endrin
Ethyl-benzene	Flouro-ethane
Halo-ethers (other)	Halo-methanes (other)
Hepta-chloro Compounds (other)	Hexachlorobutadiene
Hexachlorocyclohexane	Hexa-chloro-cyclo-pentadiene
Iso-phorone	lead
Mercury	Naphthalene
Nickel	Nitrobenzene
Nitro-phenols	Nitrosamines
Pentachlorophenol	Phenol

Phalate esters	Polychlorinated biphenyls
Polynuc. aromatic hydrocarbons	Selenium
Silver	Tetra-chloro-di-benzodioxin
Tetra-chloro-ethylene	Thallium
Toluene	Toxa-phene
Trichloroethylene	Vinyl chloride
Zinc	

ELEMENTS OF CALCULATION

Objective of the calculation is to produce reliable data, information and results in a professional manner which are reliably accurate, readily usable, easy to read and simple to retrieve and reuse. Therefore a sequential approach for a calculation, if followed, may yield in acceptable results with less complication. A sample calculation generally should have ten (10) steps or elements. They are:

Objectives

Approach

References

Symbols and Units

Assumptions

Conversion formulas

Operating and Properties Data

Calculations

Results and Summary Information

These elements would prepare to encourage use of proper software, setup for reviewing and provide assumptions, inputs, calculations, supporting information, summary information etc. However more complicated calculations may not have always these attributes.

Suggest that the calculations whether performed by experts or prepared by knowledgeable professionals should be checked by others with detailed feed back. It thus would reduce errors of omission and reinforces fresh viewpoints enhancing the calculation reliability.

COMMON CONVERSION TABLES:

Convert	Multiply By	Get
Length (Meter):		
Feet	12	Inch
Meter, m	3.3	Feet
Meter	100	Centimeter
Meter	1000	Millimeter, mm
Inch	2.54	Centimeter
Yard	3	Feet
Yard	36	Inch
Area (Square Meters):		
Square Meters, m^2	10.76	Square Feet
Square Yards	O.836	Square Meters
Acres	43560	Square Feet
Volume, Capacity (Cubic Meters):		
Cubic Meters, m^3	35.31	Cubic Feet
Cubic Meters	1.308	Cubic Yards
Cubic Yards	27	Cubic Feet
Cubic Feet	1728	Cubic Inches
Cubic Feet (Water)	7.48	Gallons

Convert	Multiply By	Get
Volume, Capacity (Cubic Meters):		
Cubic Feet (Water)	62.4	Pounds
Cubic Centimeters	0.0002642	Gallons
Milliliters	0.0002642	Gallons
Liters	0.2642	Gallons
Cubic Meters	264.2	Gallons
Grams	0.0002642	Gallons
Pounds	0.1198	Gallons

Convert	Divide By	Get
Volume, Capacity (Cubic Meters):		
Cubic Centimeters	3785	Gallons
Milliliters	3785	Gallons
Liters	3.785	Gallons
Cubic Meters	0.003785	Gallons
Grams	3785	Gallons
Pounds	8.34	Gallons

Convert	Divide By	Get
Mass (kg):		
Kilograms	0.4536	Pounds
Grams	28.35	Ounce
Grams	453.6	Pounds
Grams (multiply by)	15.43	Grain
Milligrams	64.8	Grain

Convert	Divide By	Get
Amount of Substance (mol.):		
kmol	0.4536	lb.mol
kmol	0.0446	std.cubic
meters		
		(0°C,! atm.)
Density (kg/m^3):		
Kg/m^3	16.02	lb/ft^3
Kg/m^3	1	g/L
Pressure (Pa):		
kPa	101.3	atm
kPa	100	bar
kPa	6.895	lb/in^2
kPa	3.377	in Hg (60°F)
kPA	0.2488	in H2O (60°F)
kPa	0.1333	mm Hg (0°C)
psi (multiply by)	2.31	ft H2O (60°F)

Convert	Multiply By	Get

Temperature (K, Kelvin or °C, Celsius):

Convert	Multiply By	Get
K	1.8	°R
°C	1.8	(°F - 32)
°C	1.0	(K- 273.15)
°C	1.0	(°R/1.8 - 273.15)
°F	1.0	(°R- 459.67)
°R	1.0	(°F+459.67)

Convert	Divide By	Get

Power (kW):

Convert	Divide By	Get
kW	0.7457	Horsepower
W	1.359	ft.lb/second.
W	1.162	kcal/hour.
W	0.2931	Btu/hour.
Horsepower (multiply)	33000	ft.lb/minute

Common Formulas:

Parameter	**Formula**
1. Specific Volume	Specific Volume = 1 / Density.
2. Specific Gravity	Gf for Liquids = ρ / ρo, where ρ0 = 999 kg/m³ at 15.6 °C and at atmos.pr. (1.013 barA) or 62.36 lb /ft³ at 60°F and at atmos.pr. (14.73.psiA). Gg for gases = ρ / ρo, = Mw / 28.96 at std. cond. where Mw is mol.wt. 0f gas.
3. Mass Flow Rate	Mass Flow Rate = Volume Flow Rate times Density
4. Organic Loading Rate	OLR (WWTP) = BOD / MLVSS

5. Sludge Age	Sludge Age = Solids in Aeration (MLSS) divided by Solids entering Aeration (SS in primary effluent).
6. Solids Retention Time	SRT=Solids in Aeration divided by Solids leaving Aeration.
7. Mean Cell Residence Time	MCRT = Volatile Solids in Aeration divided by Volatile Solids leaving Aeration
8. BOD loading	Kilogram per cubic meter per day= 0.0624 lb/cu ft. day
9. Horsepower (SI)	HP (SI) = 1.01387 HP (mech.) HP (mech.) = 33000 ft lb/min. BHP= ((cubic meter per hour times height (m)) Divided by (271.23 times efficiency)) times (specific gravity).

10. ppm by volume (20 ° C)
    = (mol.wt./24.04) mg per cubic meter
    = (mol.wt./0.02404) μg per cubic meter
    = (mol.wt./24.04) μg per liter
    = (mol.wt./28.8) ppm by weight
    = (mol.wt divided by(385.1 times 10 to the power 6) pounds per cu ft.

11. ppm by weight
    = 1.198 mg per cubic meter
    = 1.198 μg per liter
    = (28.8 divided by mol.wt.) ppm by vol.
    = (7.48 times 10 to the power negative 6) pounds per cu ft.

12. Btu
    = 1/180 of the amount of heat required to raise 1 pound of pure water from 32ºF to 212ºF.
    = 1055.18 joules
    = 252 calories
    = 107.6 kilogram (force) -meters
    = 1/2545 horsepower-hours
    = 1/3413 kilowatt-hours

13. Boiler Horsepower
    = 33524 Btu/hour
    = 9822 kilowatts

14. Ton Refrigeration
    = 12000 Btu/hour

EPA REGIONS AND DIRECTORY

STATE	USPS CODE	EPA REGION
Alabama	Al	4
Alaska	AK	10
Arizona	AZ	9
Arkansas	AR	6
California	CA	9
Colorado	CO	8
Connecticut	CT	1
Delaware	DE	3
Dist. of Columbia	DC	3
Florida	FL	4
Georgia	GA	4
Guam	GU	9
Hawaii	HI	9
Idaho	ID	10
Illinois	IL	5
Indiana	IN	5
Iowa	IA	7
Kansas	KS	7
Kentucky	KY	4
Louisiana	LA	6
Maine	ME	1
Maryland	MD	3
Massachusetts	MA	1
Michigan	MI	5
Minnesota	MN	5

Mississippi	MS	4
Missouri	MO	7
Montana	MT	8
Nebraska	NE	7
Nevada	NV	9
NewHampshire	NH	1
New Jersey	NJ	2
New Mexico	NM	6
New York	NY	2
North Carolina	NC	4
North Dakota	ND	8
Ohio	OH	5
Oklahoma	OK	6
Oregon	OR	10
Pennsylvania	PA	3
Puerto Rico	PR	2
Rhode Island	RI	1
South Carolina	SC	4
South Dakota	SD	8
Tennessee	TN	4
Texas	TX	6
Utah	UT	8
Vermont	VT	1
Virginia	VA	3
Virgin Island	VI	2
Washington	WA	10
West Virginia	WV	3
Wisconsin	WI	5
Wyoming	WY	8

Region	1	JFK Federal Building, Boston, MA 02203 617 565 3420
Region	2	26 Federal Plaza, New York, NY 10278 212 637 3000
Region	3	841 Chestnut Street, Philadelphia, PA 19107 215 566 5000
Region	4	245 Courtland Street, N.E, Atlanta, GA 30308 404 562 8327
Region	5	230 South Dearborn Street, Chicago, IL 60604 312 353 2000, 800 572 2515, 800 621 8431
Region	6	1201 Elm Street, Dallas, TX 75270 214 665 6444
Region	7	726 Minnesota Avenue, Kansas City, MO 66101 913 551 7000

Region	8	999 18 th Street, Suite 1300, Denver, CO 80202 303 312 6312, 800 759 4372
Region	9	215 Fremont Street, San Francisco, CA 94105 415 744 1500
Region	10	1200 Sixth Avenue, Seattle, WA 98101 206 553 1200

EPA 1200 Pennsylvania Avenue
N.W., Washington, DC 20460

800 368 5888 (Asbestos)

800 490 9198 (Clearing House)

800 535 0202 (Emergency Planning)

202 260 7751 (Information Center)

202 554 1404 (Office of Toxic Substances)

800 858 PEST (Pesticides)

800 426 4791 (Water)

202 564 5700 (Water)

800 424 9346
(Superfund—RCRA/CERCLA)

States, Counties, Cities-all have environmental and/or public health agencies. They can help and/or direct to the right agencies for compliance issues.

EPA REGULATORY TOOLS

EPA along with the states is covering to achieve, maintain and protect the environment and public through these regulatory tools.

New source review and prevention of deterioration permits

Title V operating permits (CAA)

New source performance standards

NESHAPs

RMPs

TRI and annual effluent/emission inventories

Air/water dispersion models

Trading and banking programs

Penalties and

Others

EPA at present to comply with its RMP requirements has at least 77 toxic substances and 63 flammable substances each with proper threshold quantity. EPA's RMP is similar, but entirely complimentary, to OSHA's PSM standard. A PSM facility can avoid RMP requirements, however, sometimes a non-PSM facility may have to comply with RMP regulations.

Compliance, although a must, should be viewed from cost-effectiveness, knowledge management and management of change. Development of process principles, processing technologies, instrumentation and automation including total cost of ownership, spare parts and maintenance are the key disciplines wherefrom cost effectiveness should be realized. Ongoing MACT analyses and standards need to be evaluated for compliance also.

Most of the states do the completion of permitting through the applications with EPA concurrence.

EPA recently released PASS interactive package which asks users the questions and uses the answers to fill out NPDES permit applications. NPDES general permit forms are:

Form 1 - General Information
Form 2A - Municipal Dischargers
Form 2B - Concentrated Animal feeding Operation/Production
Form 2C - Manufacturing
Form 2D - New Sources/Dischargers for Process Water
Form 2E - Facilities which do not discharge Process Water
Form 2F - Storm Water Discharge associated with Industrial Activity
Form 2S - Biosolids

A written basic compliance plan for each entity, facility or plant with objectives and targets, data (basic, design, environmental and safety) and information, and an implementation plan is required to satisfy most of the EPA compliance issues. The plan must have the following common elements program in it or as an addendum to it from a program, procedure, plan or data for the same entity, facility or plant. The common elements are:

Process and equipment information,
Employee involvement,
Worst-case scenario,
Emergency response plan,
Hazard review,
Incident investigation,
Operating and maintenance procedure,
Safe work including hot work plan,
Contractors' work procedure,
Training,
Management of Change and
Follow-up.

WEB SITE DIRECTORY

Selected Government and Professional Organizations:

ACC	www.cmahq.com
ACS	www.acs.org
ACGIH	www.acgih.org
AIChE	www.aiche.org www.aiche.org/cwrt
ANSI	www.ansi.org
API	www.api.org
ASHRAE	www.ashrae.org
ASME	www.asme.org
ASPA	www.aspa.org
ASSE	www.asse.org
ASTM	www.astm.org
ASQ	www.ASQ.org

AWMA	www.awma.org/resources
AWWA	www.awwa.org
APWA	www.pubworks.org
CDC	www.cdc.gov
CIMS	www.chemtracker.com
CSB	www.csb.gov
CSPA	www.cspa.org
CSPC	www.cspc.gov
DOE	www.doe.gov
DOE(EM)	www.em.doe.gov
DOT	www.dot.gov
EPA	www.epa.gov www.epa.gov/airnow www.epa.gov/clearinghouse www.epa.gov/empact www.epa.gov/ceampubl/index.htm www.epa.gov/tools/rmp-comp www.epa.gov/enviro/html/em/ www.epa.gov/greenchemistry www.epa.gov/iriswebp/iris/index.html

www.epa.govn/ncepihom/ordering/
www.yosemite.epa.gov/oswer
www.epa.gov/ost/beaches
www.epa.gov/ttn/atw
www.epa.gov/ttn/scram
www.epa.gov/waters
www.epa.gov/owm
www.epa.gov/ow
www.epa.epa.gov/win/
www.cfpub.epa.gov/npdes

for 40 CFR 50, 63, 68,122,141 &—other regulations

FEMA www.fema.gov

FDA www.fda.gov

HUD www.hud.gov/emaps

ISA www.isa.org

ITER www.tera.org/iter

NAS/E www.nap.edu/info/browse/htm

NFPA www.nfpa.org

NCSE www.ncseonline.org

NSC www.nsc.org

OSHA	www.osha.gov
	for 29 CFR 1910 &—regulations
SIRI	www.siri.uvm.edu/msds www.ilpi.com/msds/index.chtml
SFPE	www.sfpe.org
USGS	www.usgs.gov www.btdqs.usgs.gov/acidrain www.uswatersnews.com
WEF	www.wef.org www.weftec.org www.scorecard.org www.w-ww.com
WHO	www.who.int
WSO	www.worldsafety.org
WQA	www.wqu.org

Selected Industrial, Non-profit and Trade Organizations:

Chemical Engineering
www.che.ufl.edu/WWW.CHE/index.html

Chlorine Institute
www.cl2.com

Compressed Gas Association
www.cganet.com

Dangerous Goods Advisory Council
www.hmac.org

Fertilizer Institute
www.tfi.org

Institute of Makers of Explosives
www.ime.org

MSDS online
www.MSDSonline.com

Nat. Association. Of Chem. Distr.
www.nacd.com

SOCMA
www.socma.com

Safety-Kleen
www.safety-kleen.com

US Water News online
www.uswaternews.com

World Resources Institute
www.wri.org

Water and Wastewater Web
www.w-ww.com

OTHER HELPFUL RESOURCE INFORMATION:

Selected Nonprofit Organizations:

Am. Council on Science on Health,
1995 Broadway, New York, NY 10023;
212 362 7044

American Medical Association,
535 North Dearborn St. Chicago, IL 60610;
312 464 4818

Asbestos Information Association,
1725 J. D. Highway, Arlington, VA 22202;

Citizens for Better Environment,
59East Van Buren, Chicago, IL 60605;
312 939 1984

Consumer Federation of America,
1424 16th Street NW, Washington DC;
202 387 6121

Environmental Defense Fund,
257 Park Ave. South, New York, NY 10010;
212 505 2100

Greenpeace USA,
1436 U St. NW, Washington DC 20009;
202 462 1177

Nat. Audubon Society,
950 Third Ave., New York, NY 10022;
212 832 3200

Nat. Coalition for Clean Indoor Air,
316 Pennsylvania Ave., Washington DC;
202 797 5436

Nat. Coalition against the use of Pesticides,
530 7th Street, Washington DC;
202 543 5450

Nat. Safety Council,
1121 Spring Lake Dr., Itasca, IL 60143;
800 621 7619

Nat. Sanitation Foundation,
P.O.Box 1468, Ann Arbor, MI 48106;
313 769 8010

Nat. Wildlife Federation,
1400 16th St. NW Washington DC 20036;
202 797 6800

Nature Conservancy,
1815 N Lynn St., Arlington, VA 22209;
703 841 5300

Radon Tech. Info. Serv.,
Research Triangle Park, NC 27709;
919 541 7131

U.S. Council for Energy Awareness,
1776 I Street N.W., Washington DC;
202 293 0770

U.S. Public Interest Res. Group,
215 Pennsylvania Ave. SE, Washington DC;
202 546 9707

World Wildlife Fund,
1250 24th St. NW, Washington DC 20037;
202 293 4800

World Resources Institute,
10 G Street, NE (Suite 800),
Washington, DC 20002
202-729-7600

Selected Professional and Trade Associations:

Air and Waste management Association,
P.O. Box 2861, Pittsburgh, PA 15230;
412 232 3444

American Geological Institute,
4220 King Street, Alexandria, VA 22302;
703 379 2480

American Industrial Hygiene Association,
345 White Pond Dr., Akron OH 44320;
330 849 8888

American Institute of Chemical Engineers,
3 Park Ave., New York. NY 10016;
800 242 4363

American Institute of Chemists,
7315 Wisconsin Ave., NW, Bethesda MD 20814:
301 652 2447

American Water Works Association,
6666 W. Quincy, Denver, CO 80235;
303 794 7711

Nat. Asso. of Env. Professionals,
5165 MacArthur Blvd., NW, Washington DC;
202 966 1500

Society of Ecological Restoration,
1207 Seminole Highway, Madison, WI 53711;
608 262 9547

APPENDIX I

Further Readings:

1. Center of Chemical Process Safety, AIChE - Various Publications
2. EPA (U.S.) - Various Documents, Publications and Standards
3. Federal Register - Various Documents
4. NFPA - 70E (OSHA 1910.331-5), 496, 704 and other Publications
5. OSHA - Various Documents and Publications (DOL)
6. API and CMA - Responsible Care and Related Documents
7. AIHA - Strategies for Effective Risk Communication Documents
8. NIOSH and Risk Institute - Publications
9. ISO—9000 and 14000 guidelines
10. NRC—10 CFR—for Nuclear Power Industries (DOE)

11. ANSI/ISA—Safety Related Documents

12. IEC—Safety (Functional) Related Documents

13. United Nations Department of Public Information - Documents on water/sanitation/energy/health/agriculture/biodiversity

14. WHO - Documents on health/disease/smoking

APPENDIX II

The National Ambient Air Quality Standards (NAAQS):

The air quality has two levels of standards defined and recorded in details in 40 CFR Part 50. The primary standard has been set to protect the public health and the secondary standard has been designed to protect public welfare/environment.

Pollutant	Primary (µg/m³)	Secondary (µg/m³)
NO_x (Annual mean)	100	100
NO_2—annual mean	not more than 0.053 ppm	0.053 ppm
SO_2 (Annual mean)	80	
(1 time/year-24 hr. max.)	365	
(1 time/year-3 hr. max,)		1300

SO_2—24 hr. average	not more than 0.14 ppm	
		3 hr. average 0.50 ppm
PM_{10} (Annual mean)	50	50
(24 hr. average)	150	150
$PM_{2.5}$ (Annual mean)	15	15
(24 hr. average)	65	65
CO		
(1 time/year-8 hr. average)	10	
(1 time/year-1 hr. average)	40	
O_3 (1 time/year- 1 hr. max.)	235	235
O_3—1 hr. average	not more than 0.12 ppm	0.12 ppm
Pb (quarterly mean)	1.5	1.5

Note: ***Pollutants are: Oxides of Nitrogen and Nitrogen Oxides; Sulfur Dioxide; Particulate Matters-10 microns or smaller and 2.5 microns or smaller; Carbon Monoxide; Ozone; and Lead.***

APPENDIX III

EPA Drinking Water Standards:

The EPA sets the standards for water. The water has primary and secondary standards. Primary standard is considered for the constituents mostly having harmful effects and therefore its toxic constituents must remain below MCL and the secondary standard is more for aesthetic reasons like for *color, smell or taste of water.*

EPA may and do improve the figures from time to time due to research/availability of new data on health effect and/or for technology. Some times the recommended level for drinking water is lower than the MCL level. Recommend that EPA or local state agencies be contacted to get the latest standards.

Contaminants	MCL (ppb)	Health Effects
Microbiological (total Col.) (Coliforms—bacteria, Fecal streptococcal etc.)	1/100 milliliters	Not necessarily produce disease but can indicate levels of organisms

		causing various illnesses.
Turbidity	1-5 tu	Disinfection of water
Arsenic	50	Toxicity / nervous system
Barium	1000	Circulatory System
Cadmium	10	Kidney
Chromium	50	Liver / Kidney
Fluoride	4000	-
Lead	50	Kidney / Nervous System; very toxic to infants and pregnant women
Mercury	2	Kidney / Nervous System
Nitrate	10,000	-

Selenium	10	Gastrointestinal
Silver	50	Skin
Benzene	5	Cancer Risk
Endrin (Insecticide)	0.2	Nervous Syst.
Lindane (Insecticide)	4	Nerv.Syst./Liver
Methoxychlor (Insecticide)	100	Nerv.Syst./ kidney
Toxaphene (Insecticide)	5	Cancer Risk
Carbon Tetrachloride	5	Cancer Risk
Trichloroethylene	5	Cancer Risk
Vinyl Chloride	2	Cancer Risk
pH	6.5-8.5	-
Hardness(TDS)	Soft: Less than 17 ppm	-
	Hard (low): 17-50 ppm	-
	Hard(mod.): 50-100 ppm	-

	Hard: 100-200 ppm	-
	Very Hard:above 200 ppm	-
Chloride	250 ppm	-
Copper	1 ppm	-
Sulfates	250 ppm	-
Zinc	5 ppm	-
Iron	0.3 ppm	-
Manganese	0.05 ppm	-

APPENDIX IV

Basic Electrical (DC and AC) Power:

Direct Currents (DC) are constant in time while the magnitude of alternating currents (AC) is a sine or cosine function of time. For a component with current I through and voltage V across, the electrical power P delivered is:

$P = IV = I^2R = V^2/R$ with R as resistance.
Power unit is generally watt.

Normally wiring for household, 120 V AC, delivers 0.5 Amp. AC to a light bulb with a resistance of 240Ω and so delivers a time—averaged power of 60 W.

Electrical work is equivalent to mechanical work which is the product of force and distance.
1 watt second (W s) = 1 Joule = 1 J = 1 N m; therefore 1 watt = 1 J/s = 1(N m)/s

Wiring practices in USA for hazardous locations are based by divisions of classes and categories instead of zones mostly applied in Europe.

For Electrical Safety requirements, go through and apply ANSI C2, NFPA 70E, OSHA

29 CFR 1910.269/.331-.335 standards and other safe work practices.

APPENDIX V

Safety Integrity Levels:

Safety integrity Levels (SIL) determine abnormal/failure/reliability measures need to incorporate in the process/plant design functions including power generation plants with the help of probability from qualitative assessment. As an example, the probability of failure on demand value for SIL 4 (highest safety impact) is greater than or equal to 1 in 100000 but less than 1 in 10000.

Based on SIL, SIS can be designed to function to take the process or processes to a relatively safe state when or where the predetermined operating conditions can not be maintained. **OSHA** maintains that the standard S84 of ISA needs to apply for SIS. More comprehensive in SIL scope is IEC 61508.

SIS is the core layer of plant and environmental safety followed by relief devices, physical containment (dikes), plant emergency evacuation and community emergency response. Plant regular process control and pre-warning alarms are meant for predetermined operating conditions and are not designed to take the plant to a safe state.

APPENDIX VI

Industry-Specific Sector Notebooks:

EPA has developed industry specific sector notebooks associated with major pollution issues. It also profiles environmental and applicable federal regulations. Notebooks available include:

Agricultural Chemical, Pesticide and Fertilizer Industry;

Agricultural Livestock Production Industry;

Electronic and Computer Industry;

Fossil Fuel Electric Power generation industry;

Inorganic Chemical industry;

Iron and Steel Industry;

Lumber and Wood Products industry;

Metal Mining Industry;

Non-fuel, Non-metal Mining Industry;

Oil and Gas Industry;

Organic Chemical Industry;

Pharmaceutical Industry;

Printing Industry;

Pulp and Paper Industry;

Water Transportation Industry and

Federal/Local Government Facilities and Operation.

APPENDIX VII

Greek Alphabet:

The letters of Greek alphabet are shown here for convenience as they are used commonly in scientific and technical disciplines for abbreviations and symbols.

Alpha	Α	α	Nu	Ν	ν
Beta	Β	β	Xi	Ξ	ξ
Gamma	Γ	γ	Omicron	Ο	ο
Delta	Δ	δ	Pi	Π	π
Epsilon	Ε	ε	Rho	Ρ	ρ
Zeta	Ζ	ζ	Sigma	Σ	σ
Eta	Η	η	Tau	Τ	τ
Theta	Θ	θ	Upsilon	Υ	υ
Iota	Ι	ι	Phi	Φ	φ
Kappa	Κ	κ	Chi	Χ	χ
Lambda	Λ	λ	Psi	Ψ	ψ
Mu	Μ	μ	Omega	Ω	ω

APPENDIX VIII

Important Hotline 800 Numbers:

CMA	Chemical Transportation Emergency Center 800 424 9300
MSDS Information	800 262 8200
DOE	Energy Efficiency Clearing House 800 363 3732
DOT	Hazardous Material Information Center 800 467 4922
EPA	Asbestos 800 368 5888
Air Quality	800 438 4318
Pesticides	800 858 7378
National Response Center	800 424 8802
Environmental Justice	800 962 6215

Environmental Publication	800 490 9198
RCRA	800 535 0202
Drinking Water	800 426 4791
Wetlands	800 832 7828
Information Center	800 887 6063
OSHA	Information Center 800 321 6742
NIOSH	800 356 4674

Note: ***All information was correct and true at press time. However these may change without notice.***

ABOUT THE AUTHOR

Satya R. Chatterjee, PhD, PE, is an independent consultant for environmental and management services. He is also an author and a speaker, not only on engineering and environmental issues but also on diverse topics like ecology, religion, philosophy, law and governance.

He retired from industry after almost four decades of service to Petrochemical, Petroleum, Oil & Gas and Engineering & Construction industries.

Besides holding a doctoral degree, he is a registered professional engineer, a member of AIChE, a member of ISA and a member of WEF. He is also a life member of Asiatic Society of India, one of the oldest illustrious learned institutions in the world.

Amongst the companies he worked for are Aramco, Bechtel and Exxon etc.

The pocketbook is a result of his keen and impressive knowledge and experience in the process, engineering, environment, construction and other related disciplines as an engineer, a management executive or a project leader.

Chatterjee, his wife and their children (all grown up) have been and are involved in many voluntary spiritual, social, educational and cultural services. Due to their initiative, effort and sponsorship, several robust service organizations have been established and are sustained.

www.ingramcontent.com/pod-product-compliance
Ingram Content Group UK Ltd.
Pitfield, Milton Keynes, MK11 3LW, UK
UKHW041821200726
13854UKWH00001BA/256

9 781414 003009